ADDITIONAL RECORDS AND EXTENSIONS OF KNOWN RANGES OF MAMMALS FROM UTAH

STEPHEN DAVID DURRANT
RICHARD M. HANSEN
M. RAYMOND LEE

ZINC READ

Published by
This edition copyright © 2024 by Zinc Read
Design and setting By Zinc Read
Email- zincread@gmail.com
ISBN-978-93-6277-407-1

CONTENTS

ADDITIONAL RECORDS AND EXTENSIONS OF KNOWN RANGES OF MAMMALS FROM UTAH

The Museum of Zoology, University of Utah, contains approximately 5000 specimens in addition to those available to Durrant (1952) when he prepared his account of the "Mammals of Utah, *Taxonomy and Distribution*." Study of this material discloses two kinds of mammals not heretofore known to occur in Utah, and extends the known limits of occurrence of many others as is set forth below in what may be thought of as a supplement to the aforementioned report of 1952.

Our study was financed in part by a grant from the National Science Foundation.

Sorex vagrans obscurus Merriam. Dusky Shrew.--Twelve specimens are available from the Abajo Mountains and Elk Ridge, San Juan County, Utah, as follows: North Creek, 6 mi. W Monticello, 8300 ft.; 1 mi. S Twin Peaks, 9500 ft.; Kigalia R. S., 8000 ft.; and Gooseberry R. S., 8250 ft. Previously, the only known specimens from east of the Colorado River in Utah were from the La Sal Mountains in extreme eastern Grand County and extreme northern San Juan County. These twelve specimens extend the known area of occurrence of the species in Utah approximately 80 miles to the south, and indicate that this shrew occurs throughout the state in favorable habitats.

Sorex palustris navigator (Baird). Water Shrew.--An individual was observed by M. Raymond Lee at North Creek, seven miles west of Monticello, Abajo Mountains, 8000 feet, San Juan County, on July 10, 1954. Usually we are extremely reluctant to record sight records, but do so in this instance because the water shrew is so distinctive that it can be readily recognized and because the occurrence extends the known range approximately 80 miles southward in Utah. This individual was observed at close range while swimming and foraging in North Creek, and there can be no doubt of its identity.

Myotis yumanensis yumanensis (H. Allen). Yuma Myotis.--Durrant (1952:43) reported this subspecies from Utah on the authority of Hardy (1941:289) who had specimens from two localities in extreme southwestern Utah. Durrant (1952:41) referred specimen Number 6784, from Willow Creek, 25 miles south of Ouray, Uintah County, to *Myotis lucifugus carissima*. Restudy of this specimen reveals that it is *Myotis yumanensis yumanensis*. This identification has been corroborated by Dr. Philip H. Krutzsch, University of Pittsburgh, Pennsylvania, and extends the known range of the subspecies *M. y. yumanensis* approximately

300 miles northeastward in Utah. See also Krutzsch and Heppenstall (1955:126) who record specimens from 2 mi. SW Jensen.

Myotis subulatus melanorhinus (Merriam). Small-footed Myotis.--This bat previously was known from only seven localities in Utah, which indicated that it occurred in only the western and southern areas of the state. Four additional records are now available from the following localities: Logan Canyon Cave, 15 miles north of Logan, Cache County; Weber College Campus, Ogden, Weber County; University of Utah Campus, Salt Lake City, Salt Lake County; Six Mile Canyon, 3-1/2 miles east of Sterling, Sanpete County. These occurrences extend the known range to the eastward in Utah, and indicate a state-wide distribution. Specimens of the subspecies *Myotis s. melanorhinus* are recorded also from as far north as Double Springs, Custer County, Idaho (Davis, 1939:117).

Pipestrellus hesperus hesperus (H. Allen). Western Pipistrelle.--Heretofore, the northernmost known specimens of this bat from Utah were from Old Lincoln Highway, 18 miles southwest of Orr's Ranch, Tooele County. Specimen Number 7531 is now available from cliffs NE [3 mi.] Ogden, Weber County, and extends the known range of this species in Utah approximately 100 miles northeastward. This pipistrelle probably inhabits all of northern and northwestern Utah in suitable habitats. This probability is supported by Davis' (1939:120) report of a specimen from Salmon Creek, eight miles west of Rogerson, Twin Falls County, Idaho.

See also Krutzsch and Heppenstall (1955:127) who record a specimen from, eastern Utah as far north as, Desert Springs which is 10 mi. SW Ouray, Uintah County.

Corynorhinus rafinesquii pallescens Miller. Long-eared Bat.--Formerly, the northernmost record of the long-eared bat in Utah was from east of Springville, Utah County. Specimens are now available from Goldhill, Tooele County, and from South Fork, Ogden River, Weber County. Professor J. S. Stanford, Department of Zoology, Utah State Agricultural College, informed us (by letter) that this bat is the common cave bat in Logan Canyon, Cache County. This northern extension of known area of occurrence of approximately 100 miles indicates that it probably is state-wide in distribution in suitable habitats. It can be inferred from Hall (1946:161) that the range of *C. r. intermedius* in Nevada extended northeastward into northwestern Utah, and Davis (1939:124) reported specimens from Bingham and Bannock counties, Idaho, that he referred to the above mentioned subspecies. This led Stanford to comment (*in litt.*) that bats of this species from northern Utah in Cache County might be *C. r. intermedius*. Insofar as we are aware, *C. r. pallescens* differs from *C. r. intermedius* only in being slightly paler. Our specimens from Goldhill and South Fork of the Ogden River are not beyond the range of color of specimens from elsewhere in the state that are referable to *C. r. pallescens*. Inasmuch as specimens are not available from Logan Canyon, we deem it best pending the acquisition of specimens from that locality to refer all members of this species from Utah to the subspecies *C. r. pallescens*.

Marmota flaviventer nosophora Howell. Yellow-bellied Marmot.--Durrant (1952:101) did not indicate that any species of the genus *Marmota* occurred on the mountains within the basin of Pleistocene Lake Bonneville. Furthermore, he commented (*op. cit.*:502) upon the dearth of sciurids within this basin. One specimen, No. 10,905, of the subspecies *M. f. nosophora* has been taken from South Willow Canyon, 10,000 feet, base of Deseret Peak, Stansbury Mountains, Tooele County. This specimen is noteworthy not only in that it extends the known range of this kind of mammal 50 miles to the west in Utah, but in that it is well within the basin of the ancient lake. The marmot is common in the Wasatch Mountains on the eastern mainland of Lake Bonneville, but to date has not been found on the Oquirrh Mountains immediately to the west. The Oquirrh Mountains are interposed between the Stansbury and Wasatch mountains. The presence of the marmot on the Stansbury Mountains indicates that it probably occurs also on the Oquirrh Mountains.

Citellus beldingi crebrus Hall. Belding Ground Squirrel.--Durrant (1952:113) had only two specimens of this ground squirrel from Standrod, Boxelder County. Additional specimens have been obtained from the following localities in northwestern Boxelder County: Grouse Creek, Park Valley, Grouse Creek Mountains, 12 miles northwest of Grouse Creek, and Goose Creek. C.

b. crebrus now is known to inhabit all the major drainages of the Raft River, Goose Creek, and Grouse Creek mountains. In addition to extending the known area of occurrence of this animal in Utah, these specimens prove also that this species is not restricted to the Snake River Drainage as Durrant (1952:113) supposed, but occurs also in the Great Basin Drainage.

Citellus richardsonii elegans (Kennicott). Richardson Ground Squirrel.--Recently, Hansen (1953:132) reported on specimens of this species from Rich and Summit counties. Additional specimens are now available from Highway 165 [2 mi. E Summit--Daggett Co. Line], 2 miles south of Utah-Wyoming State Line; 5 miles west of Manila, and one mile northeast of Manila (Carnegie Museum). These localities are in Daggett County. The occurrence of these ground squirrels in Rich, Summit and Daggett counties suggests that they occur along the entire northern piedmont of the Uinta Mountains.

Citellus lateralis trepidus (Taylor). Golden-mantled Ground Squirrel.--Durrant (1952:126) estimated that practically all of the area in Utah that is within the Great Basin might be included in the range of this subspecies. Actually, he had specimens from only the Raft River Mountains in northwestern Boxelder County. He included sight records from the Deep Creek Mountains and from the Oquirrh Mountains. Subsequently two specimens, numbers 7469A and 7470A, were obtained from the Deep Creek Mountains. To date neither specimens nor subsequent sight records have been obtained from the Oquirrh Mountains, and we are of the opinion that Durrant erred, and that the golden-mantled ground squirrel does not inhabit these mountains. Therefore, it seems at this writing that this subspecies, in Utah, occurs only in the extreme western and northwestern parts of the state in the Deep Creek and Raft River mountains, and not within the basin of Pleistocene Lake Bonneville as formerly supposed.

Eutamias umbrinus umbrinus (J. A. Allen). Uinta Chipmunk.--Two specimens, numbers 10,236 and

10,237, from the junction of Argyle and Minnie Maud creeks, Carbon County, prove that members of this subspecies occur on the West Tavaputs Plateau, which is outside the range ascribed to this subspecies by White (1953:572) and by Durrant (1952:142). The grayish color of these specimens and the reduction of ochraceous pigments constitute basis for referring the specimens to *E. u. umbrinus,* and not to *E. u. adsitus* or *E. u. montanus.* *E. u. umbrinus* on the West Tavaputs Plateau is separated from *E. u. montanus* on the East Tavaputs Plateau by the Green River and its deep chasm.

Perognathus formosus incolatus Hall. Long-tailed Pocket Mouse.--Prior to the description of this subspecies by Hall (1941:56), animals of this species had not been reported from within the basin of the Pleistocene Lake Bonneville. When Durrant (1952) prepared his manuscript he had but a single specimen from western Millard County and one nearby record (Fautin, 1946:280). Additional specimens are known from the following localities: junction of Trout Creek and Birch Creek, Deep Creek Mountains, Tooele County; six miles north of Ibapah, Tooele County; five miles south of Timpie, Tooele County; north end of Newfoundland Mountains, Boxelder County; and Groome, Boxelder County. These occurrences show that the species is not restricted to the extreme western part of the state, but occurs in suitable habitats throughout the basin of Pleistocene Lake Bonneville. The known range is extended approximately 150 miles north and 40 miles east.

Perognathus parvus trumbullensis Benson. Great Basin Pocket Mouse.--Durrant (1952:477), suspecting that this mouse occurred in Utah, included the subspecies *P. p. trumbullensis* in his hypothetical list. Numerous specimens are now available from the following localities: Pine Valley Mountains, Enterprise Reservoir, and 19 miles west of Enterprise, Washington County; Bown's Reservoir, Snow Ranch, Hall Ranch,

Steep Creek, Garfield County; Aquarius Guard Station, Aquarius Plateau, Wayne County. Insofar as we are aware, these occurrences are the first to be recorded from Utah, and extend the known range of this subspecies 150 miles northward.

The specimens from Washington County are paler than those from Garfield County, and this pallor indicates intergradation with the subspecies *P. p. olivaceus*. Of animals from the Aquarius Plateau, those from the eastern and southern localities are pale and have a marked suffusion of ochraceous in the upper parts, whereas those from the western and northern localities are extremely dark owing to a heavy suffusion of black in the upper parts. The skulls of animals from the Aquarius Plateau resemble those of *P. p.*

trumbullensis in the majority of diagnostic characters. In some few characters, nevertheless, the skulls resemble those of *P. p. olivaceus,* and in other characters are intermediate between these two named subspecies. In shape and size of the interparietal, in slightly longer nasals, and in slightly greater alveolar length of upper molariform teeth, animals from the Aquarius Plateau differ from either of the aforementioned subspecies. All characters considered, we deem it best to refer these specimens to the subspecies *P. p. trumbullensis.*

Thomomys talpoides bridgeri Merriam. Northern Pocket Gopher.--An adult female, Catalogue No. 25667 of the Museum of Natural History of the University of Kansas, skin with skull, was trapped, on 30 June 1948, 14 miles south and 2 miles east of Robertson, 9,300 feet, in Summit County, Utah, by James O. Lonnquist (original number 146). This is the first record of this subspecies from Utah, and raises to 37 the named kinds of pocket gophers known from Utah.

Reithrodontomys megalotis megalotis (Baird). Western Harvest Mouse.--Durrant (1952:295) reported no harvest mice from the Uinta Basin in northeastern Utah. One specimen, No. 10,239, was obtained from two miles east of Duchesne, Duchesne County. This specimen extends the known range in Utah 50 miles northward, and indicates that the harvest mouse of the subspecies *R. m. megalotis* occurs throughout the Uinta Basin.

Onychomys leucogaster pallescens Merriam. Northern Grasshopper Mouse.--Hansen obtained specimens (in alcohol) from Kennedys Hole, junction of the White and Green rivers, Uintah County. The northernmost specimens available to Durrant (1952:328) were from one mile east of Greenriver, Grand County. These specimens from Uintah County extend the known range 80 miles to the north, and substantiate Durrant's conclusion that this subspecies occurs east of the Green and Colorado rivers.

Clethrionomys gapperi uintaensis Doutt. Red-backed Mouse.--Previously, the red-backed mouse in Utah was known only from the Uinta and Wasatch mountains. The southernmost localities from which specimens were available were in northern Wasatch County and southern Salt Lake County. Durrant (1952:355) supposed that the species ranged southward to Mount

Timpanogos in Utah County. One specimen, No. 10,075, from the summit, 18 miles east of Mayfield, Sanpete County, and 4 from Ephraim Canyon, 15 miles east of Ephraim, Sanpete County, show that this subspecies occurs also on the Wasatch Plateau of central Utah. These latter specimens extend the known range of the red-backed mouse in Utah approximately 100 miles southward. Owing to the practically continuous nature of the central mountain ranges of Utah, students of mammals of Utah usually suspect that most montane mammals occur throughout these mountain ranges. The red-backed mouse has been sought for in vain in the mountains south of the Wasatch Plateau. Suitable habitats for this mouse occur throughout the Fishlake Mountains, Thousand Lake Mountains and the

Aquarius Plateau, but despite intensive collecting, none has been obtained from these areas.

Phenacomys intermedius intermedius Merriam. Heather Vole.--The heather vole, while not rare, is uncommon in Utah. Durrant (1952:360) had but eight specimens from various localities in Summit, Wasatch, Salt Lake and Utah counties and supposed that the species was restricted to the western Uinta Mountains and southern Wasatch Mountains. In the summers of 1952 and 1953, intensive collecting of mammals was carried out on Boulder Mountain and the Aquarius Plateau, in Wayne and Garfield counties. Two specimens, nos. 8956 and 9074, were obtained from Spectacle Lake, Boulder Mountain, Garfield County. These specimens extend the known area of occurrence

175 miles southward in Utah. No specimens are known from the areas between Mount Timpanogos in Utah County, and Boulder Mountain in Garfield County. We suspect, however, that when this intervening area has been thoroughly studied, the heather vole, like many other montane mammals, will be found throughout the entire length of the central mountain ranges.

Microtus pennsylvanicus modestus (Baird). Pennsylvanian Meadow Mouse.--In Utah this mouse was known only from wet meadows in valleys immediately west of the Wasatch Mountains, as far south as a place 2 miles south of Provo (Hall and Cockrum, 1953:410). Norman V. Chamberlain collected several specimens "near" Koosharem Reservoir, Sevier County. These extend the known range of this subspecies 110 miles southward, and suggest that Pennsylvanian meadow mice occur, in suitable habitat, all along the eastern margin of the Great Basin in Utah, at least as far south as Sevier County. All northern specimens are from the drainages of Utah Lake and Great Salt Lake, but these specimens from Sevier County are from the Sevier River Drainage. This species requires a fairly moist environment, and such habitat exists between the aforementioned drainages which are practically interconnected by Mona Reservoir and its adjacent areas of springs.

Zapus princeps utahensis Hall. Big Jumping Mouse.--Durrant (1952:388) reported a specimen from Puffer Lake, Beaver Mountains, Beaver County. He supposed that this mouse occurred also at high elevations still farther south in Utah. Subsequently, two specimens were obtained from Garfield County; No. 9006 from Steep Creek, 12 miles north of Boulder and No. 9071 from East Fork of Boulder Creek, 10 miles north of Boulder. These two specimens extend the known range of jumping mice in Utah 75 miles southeastward. Several specimens have been obtained also from the Fishlake Plateau, and further bear out Durrant's supposition that these mammals occur on all of the high mountains of central Utah, at least as far south as the Aquarius Plateau.

Urocyon cinereoargenteus scottii Mearns. Gray Fox.--One skull, No. 10,240, from mouth of Birch Creek, Deep Creek Mountains, Juab County, extends the known geographic range 50 miles east from Cherry Creek Canyon, Nevada (see Hall, 1946:241). This record indicates that the species occurs in the mountainous areas on the western margins of Pleistocene Lake Bonneville, and extends the known range in Utah approximately 150 miles northward. Furthermore, this record proves that the gray fox occurs as far north in western Utah as it does in eastern Nevada, but to date none has been obtained from the mountains within the basin of the ancient lake, even though some of them are not far removed from the Deep Creek Mountains.

Lutra canadensis nexa Goldman. River Otter.--River otters are rare in a semi-arid state like Utah, and few have been preserved as scientific specimens. Durrant (1952:436) had access to but one skull from an immature animal from the Raft River Mountain area in northwestern Boxelder County. At present there are two complete specimens (skins, skulls and skeletons) in the collection of the University of Utah. They were trapped by an employee of the Utah State Fish and Game Department, and were donated to the University of Utah by J. Perry Egan, Director of the above mentioned department. They are nos. 8854 and 8855, and are from the Raft River, 2 miles south of the Utah-Idaho border, Boxelder County.

Alces americanus shirasi Nelson. Moose.--The moose is rare in Utah, and to date records of its occurrence have been based solely upon sight records. There are, nevertheless, two specimens preserved. One is a young bull (skull only) from Farmington Canyon, Davis County, in the collection of Weber College, Ogden, Utah. The other is one antler (No. 10,745) of a young bull from Henrys Fork, 16 miles south of the Utah-Wyoming border, Summit County, and it is in the collection of the University of Utah. This large cervid apparently is increasing in numbers in the state. Dale Jones of the Utah State Fish and Game Department reported to us that a herd of 25 animals was observed in 1954, in the vicinity of Haydens Peak, Bear River Drainage, Summit County. A cow and a calf were seen in the vicinity of Strawberry Reservoir, Wasatch County, in 1951, by employees of the same department. This latter locality is the most southern and eastern point of their known occurrence in Utah.

Ovis canadensis canadensis Shaw. Mountain Sheep.--Formerly, the mountain sheep was not known to occur in the La Sal Mountains in Grand and San Juan counties. On October 23, 1954, a two year old ram, No. 10,906, was killed by a deer hunter at a locality 1-1/2 miles north of La Sal, La Sal Mountains, San Juan County. This constitutes the first complete specimen (skin and skull) of a mountain sheep from Utah.

According to Harold Crane, of the Utah State Fish and Game Department, this ram was running with a herd of mule deer, and was the only mountain sheep that was seen. The ram was confiscated and given to the Department of Zoology, University of Utah, for preservation as a scientific specimen.

THE END